Understanding...

HOLY TRINITY
&
INVERTED HOLY TRINITY

Applied Math to Life, Research

Rui M.F. Nascimento

ISBN-13: 978-1477646649
ISBN-10: 1477646647

TABLE OF CONTENTS

HOLY TRINITY'S MEANING

HOLY TRINITY'S MEANING
(Mathew 28:19)

(Represents God, Energy and Matter, all it takes to form the Universe)

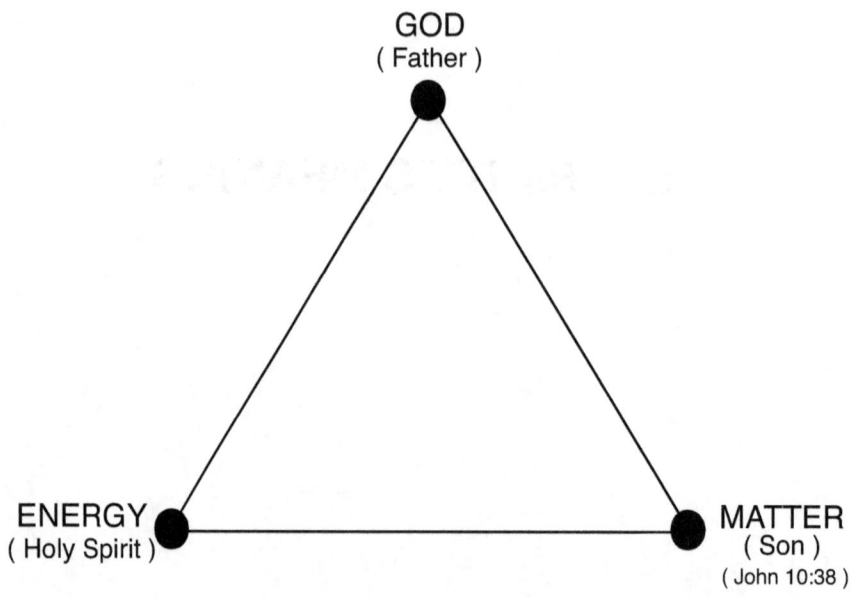

GOD
(Father)

ENERGY
(Holy Spirit)

MATTER
(Son)
(John 10:38)

Sources:

✓Allan Kardec
✓Freemasonry   All combined together.
✓Catholic Religion (Mathew 28:19)

✓"God, Father" stands above all of us, at the maximum point of evolution there is, being neutral between left and right wings. He is no one man, but a degree of evolution of the Soul, the last one in the Infinite Scale of Evolution of the Universe.

✓All the Matter in the Universe is theoretically represented at Right Wing. Human, animal, vegetable and extraterrestrial bodies, rocks, planets, meteors, chairs, tables, bricks, walls, metals, cloth, money, assets, all the tangible bodies in the Universe. The "Son" is the Matter, as the identity of "The Son", Jesus Christ or another culture's one, is attached to the Body/Matter, since its birth until its death. Spirit does not get born nor dies, does not reproduce itself as it is not sexual, only matter is. A Human Being gets its identity with its new baby body. This identity looses all its legal value and rights as the body dies, so the identity is directly related to the Body/Matter. The Spirit/Energy never dies, lives its eternal life. "The Son/Matter stands at his God/Father's Right Wing". Matter stands at Universe's Right Wing.

✓All the Energy in the Universe is Left Wing. Solar, wind, electrical, nuclear, human, animal, vegetable, spiritual, sexual, etc. The "Holy Spirit" is a type of Energy in the Universe. All the spirits are Holy, as we are all part of God, Allah, Dja, Supreme Intelligence, The Great Architect of the Universe, whatever, Master Piece of Work. Being Holy is part of their nature, what differentiates them among each others, is their degree in the Scale of Evolution. Spirits are immaterial, intangible; they are Universe's Ethereal Energy and living creatures's inner Essence. Since "Up" and "Right Wing" are already occupied, Energy can only be "Left Wing". Attention not to misjudge the Energy with the Matter used to carry it. Energy stands at Universe's Left Wing.

Being left wing, middle of the road or right wing can be expressed in a percentage of ignorance/knowledge, of energy/matter. All is relative towards something.

Like the Math student begins by learning only positive numbers, I will also start by explaining the "positive" Holy Trinity. Like in Math, the lowest of the positive numbers is 1, so in "positive" Holy Trinity, the lowest point possible, the base, is the largest part of the figure.

DISSECTING HOLY TRINITY

DISSECTING HOLY TRINITY

Everything and everyone can be explained and understood via Holy Trinity.

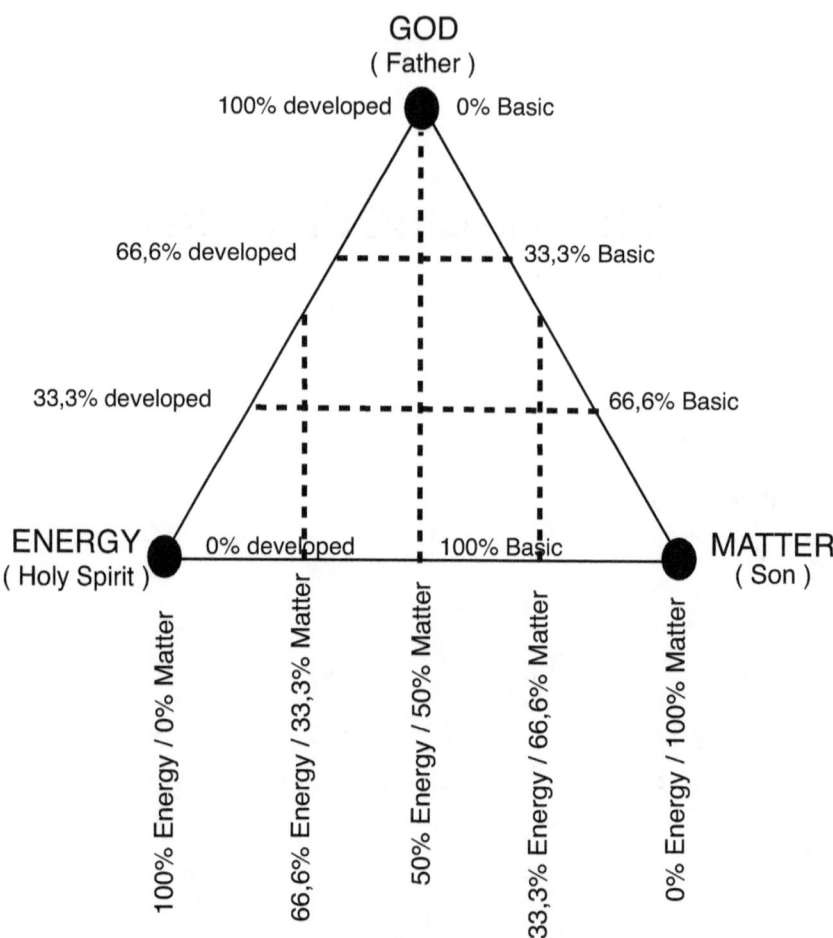

HOLY TRINITY'S CHARACTERISTICS

HOLY TRINITY'S CHARACTERISTICS
Upper Degree of Evolution (+∞)

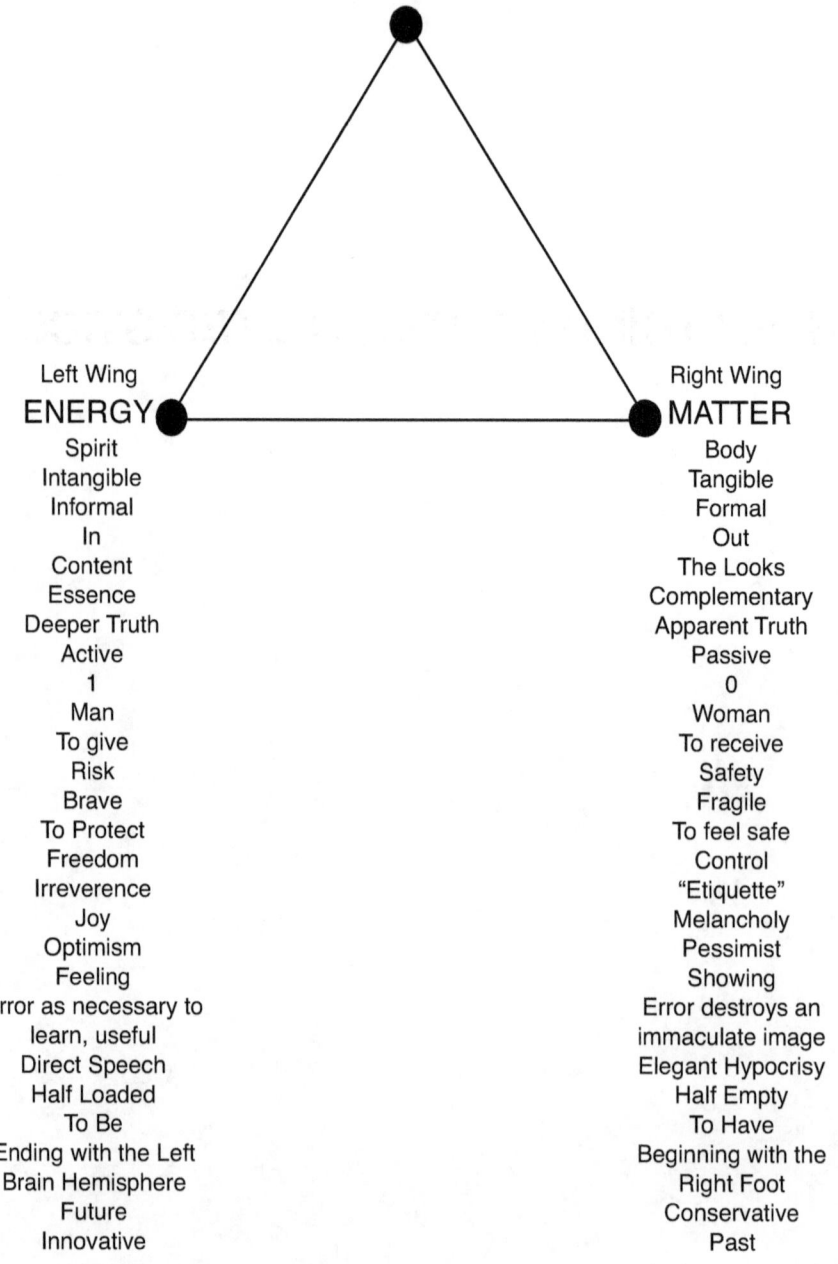

Left Wing ENERGY	Right Wing MATTER
Spirit	Body
Intangible	Tangible
Informal	Formal
In	Out
Content	The Looks
Essence	Complementary
Deeper Truth	Apparent Truth
Active	Passive
1	0
Man	Woman
To give	To receive
Risk	Safety
Brave	Fragile
To Protect	To feel safe
Freedom	Control
Irreverence	"Etiquette"
Joy	Melancholy
Optimism	Pessimist
Feeling	Showing
Error as necessary to learn, useful	Error destroys an immaculate image
Direct Speech	Elegant Hypocrisy
Half Loaded	Half Empty
To Be	To Have
Ending with the Left Brain Hemisphere	Beginning with the Right Foot
Future	Conservative
Innovative	Past
...	...

We have already realized how Energy stands on Holy Trinity's Left Wing, as Matter is Holy Trinity's Right Wing. So, Human's deeper Energy relies in its spirit, it's the human essence, its content, its human's deepest truth. The mind is used as an Interface to interact with the material body. All the Energy in the Universe, solar, wind, electrical, nuclear, human, sexual, etc., is intangible, meaning, has no form nor shape. So does Left Wing Humanity. It always has that relaxed, chilled out, informal attitude, because energy has no shape, is informal, it floats like the wind. The material human body is Right Wing, as it is Human Being's shelter, shell, it has a form that will be attached to its Spirit for an all lifetime. We cannot touch or grab a spirit, but we can touch or grab its body because Matter has always a shape, it is always formal. So it is human right wing: always very straight, very hard, very formal. But the smallest part of matter, the atom, has space in between filled with energy, positive when near protons, neutral or negative when near electrons. So does the right wing humanity. Inside all that cold blood rigidity, straightness and inflexibility, there's always a space for a warm and comforting inner human energy. Having an inner Spirit that brings life to its body shell, makes Left Wing the active part and Right Wing's body the passive part of Human Being. Like the Spirit is active by penetrating the body and giving life to it, also man is the active part by penetrating woman and giving her life, through his sperm, witch is the seed of life that will grow and develop in woman's egg. Applying the Mathematical Logical theory, Inner Truth is represented in Binary language as "1", as the Apparent Truth is represented by "0"; The Active Part is represented by "1", as the Passive Part is represented by "0". This makes perfect sense, since a number is most valued having the "1" at its Left and the "0" at its Right. For instance: "1.000.000" is worth a lot more than "0.000.001". Everyone of us would rather have a number like "1.000.000" than "0.000.001". Neither "1" or "0" are right or wrong numbers, what is right or wrong is their position relatively to each other, meaning the "1" place is at Left Wing, as "0'" place is at Right Wing. Any house, corporate or government manager, who gets presented with "0.000.001", "0.000.010" or "0.000.100" on his results, knows he is in a bad situation, that those numbers are bad. On the contrary, every house, corporate or government manager dreams and works to have "0.010.000", "0.100.000" or "1.000.000" on his results, because he knows perfectly well "1" place is at Left Wing, and "0"' place is at Right Wing. This means the cipher at the Left of the Number, rules, Meaning Left Wing Rules, meaning, theoretically, the Leader is the Active Part, the Number One, and this can only rule when he stands at Left Wing. When standing at "0"'s Right Wing, meaning at Society's Right Wing, not only that cipher "1" is "wrongly" positioned, far away from his Natural Habitat, unhappy, as all the Number, meaning All the Society it belongs to, is worth a lot less, because its real value stands too far away from its potential value.

Another way of seeing it, is using the Deeper/Apparent Truth perspective. Who do we want to rule our homes, corporations, our governments? Deeper Truth or Apparent Truth? Most of us want Deeper Truth to rule our lives and societies, meaning, the majority of us want Left Wing to Rule. What happens when Apparent Truth rules? We do have several examples of situations like that in our western World... Enron, for example, as well as many other corporations have given Apparent Truth the power to rule them through accounting artificial mechanisms. Apparent Truth brought bankruptcy, unemployment, poverty and shame to those numbers, meaning to those corporations, as well as it made perfectly clear the need for a "revolution" in that sector, giving Deeper Truth higher power to control Apparent Truth. To give Deeper Truth higher power to control Apparent Truth, turning Deeper Truth the leader over Apparent Truth. "1"' Natural habitat is Left Wing, as well as the leader position in a number. Where cipher "1" was allowed to rule over cipher "0", the market has regained confidence in the accounting procedures of the major corporations and order was restored to satisfying levels.

Being the Active part and spiritually related, Man, at the Left Wing is the one who's supposed to take the initiative, to be brave and to protect. To balance the equation, the Woman, at the passive Right Wing is supposed to return all it receives, by giving Left Wing all the necessary support, to pay off the bravery and the risks taken. This happens not only with Human Life, but in all the Universe, as this is a Holy Trinity Law, meaning, a Universal Law. The active Left Wing has all the Energy it takes to take risks, to be brave, to protect, as Right Wing feels so fragile and has this need to feel secure, protected. Having already proved its own bravery, Left Wing is self confident, leading it to be irreverent, joyful and heading straight to the future. Feeling so fragile and needing so much to be protected, Right Wing knows its fragility, so, it does not like to show how it feels deep inside, hiding it most of the times, using apparent and formal truth as its best defence. "Etiquette" suits Right Wing just fine, by giving the family, the company, friends and society in general, the image Right Wing feels and behaves the way it should. Assuming a symmetrical point of view from Left Wing, Right Wing tends to be melancholic, pessimist when regarding the future, having this repetitive tendency to talk about the past, to spend a lot of time talking about it, as well as, kind of wishing it to return, although it knows that's really not the best thing to happen. Right Wing has this fear of letting others sense it's fragility, so, most of the times, chooses to put an outside social mask, with the single purpose of pretending, acting, hiding its Energy is only 30%, 40% of its being, as the Matter takes over the remaining 60%, 70%. Assuming a symmetrical point of view, Left Wing knows its inner strength, the same way it knows its external fragile image, so whenever family, co-workers, society in general pays too much attention to its outside image and not the necessary attention to its inner capabilities, Left Wing just feels this inner demanding to be itself, whatever others may think, whatever is going to be the subjective present interpretation of Society. When Left Wing feels its inner voice, takes the initiative, knowing in the present time it is going to be hardly understood by the Society, but in the future they will gain that capacity to finally understand it, its a matter of time. Left Wing is pioneer by nature, accepting being misjudged in the present time, with this inner feeling that in the future its value will be publicly recognised; not being itself means, for Left Wing, being publicly judged by its weak image in the present, in the close future, as well as in the long future, forever... For All Eternity! Even though we can believe that luck will give us a better long future, that in the next lifetime things may be different, the Graphic of Life tells us the Universe does not really work that way... we must take each step at a time, so, expecting to appear in a higher Degree of Evolution in a future life, just by magic, is not what the active, energetic and brave Left Wing was programmed for, by (+∞, +∞), God, Allah, Odin, The Great Architect of the Universe, The Supreme Intelligence, whatever. It's just against the Holy Trinity's and the Universal Law, that specific procedure of Left Wing will always be interpreted as Unconstitutional by Inverted Holy Trinity based societies, although the only path allowed by Holy Trinity ones. By having in Apparent Truth its natural habitat, Right Wing has a major difficulty to deal with error, its own errors as well as others' errors. Right Wing feels errors destroy the immaculate image they want to pass and tear apart a social mask that took so many years and effort to turn credible to its family, its corporation, its society. The Apparent Truth of that mask can be "well" used by constructing some project, or used "wrongly" by destroying someone else's project. They are not actively aggressive because they have less Energy than Matter. With that Matter (money) they pay Left Wing services of someone who has a lot of Energy but does not have Matter (money) to support its family. Because Right Wing is low on Energy, tends to be afraid of not accomplishing its goals, tends to be shy and its lack of confidence turns out to be a natural consequence. Right Wing has always a lot more to do with the material wealthiness than for goals it manages to achieve on its own, by spending its own Energy.

Once you have learned the "positive" Holy Trinity characteristics, it's now time to learn about the Inverted Holy Trinity. Basically it's quite the same, except for being headed to (-∞, -∞) and over its greater influence. This causes Inverted Holy Trinity Souls to see All the Universe upside down, and to misjudge Left and Right Wings. For instance, Inverted Holy Trinity Souls are always trying life to be like it was in the past, instead of visioning the future, are pessimist because they have this "half empty" vision of it All, instead of being optimist by assuming a "half loaded" attitude. They see "1" at Right Wing and "0" at Left Wing, they figure Man at Right Wing and Woman at Left Wing, they switch all Left and Right Wing characteristics, because that is what every one does when is not headed to $(+\infty, +\infty)$, but to $(-\infty, -\infty)$.

Check it out...

(INVERTED) & HOLY TRINITY

HOLY TRINITY

(John 14:1-3; Psalm 23:6; Proverbs 1:20-23; 2;3;4;8)

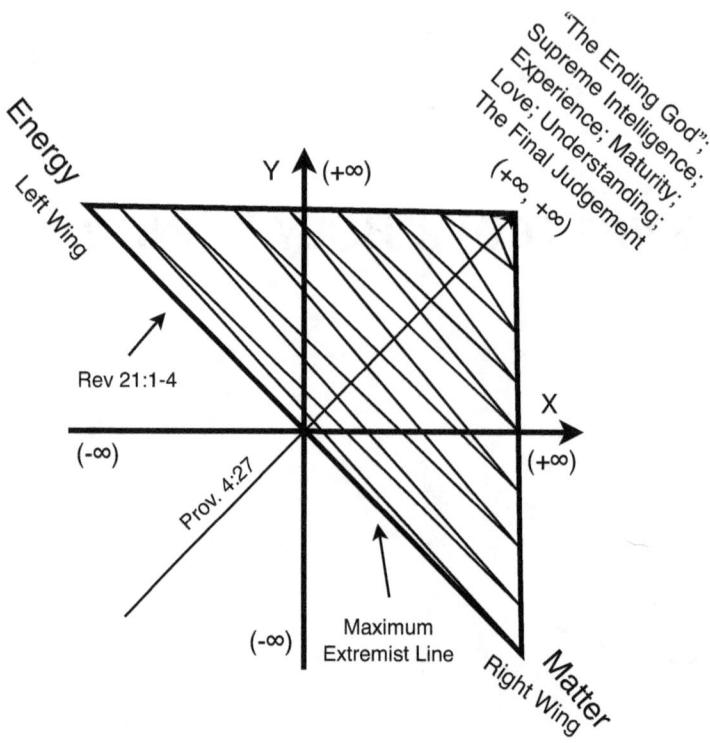

Headed towards the future, to (+∞, +∞), and over its greater influence, represents total knowledge (intellectual, emotional, spiritual, etc.), and experience. The upper world, where understanding is a must, visionaries and pioneers. "The kingdom of God", according to Jesus Christ's metaphors. For us, the beginning is already known, it's clear to us, so it's white. The future is unknown, the light of knowledge has not yet been over it, so the future is dark. Future will only stop being dark the day it will become Present, and perfectly clear (white) once it became Past.

As pioneers and innovators, we stand for the Dark, we must not fear it, actually we feel quite attracted to it. That makes us the Holy Trinity Innovative Dark Forces.

INVERTED HOLY TRINITY

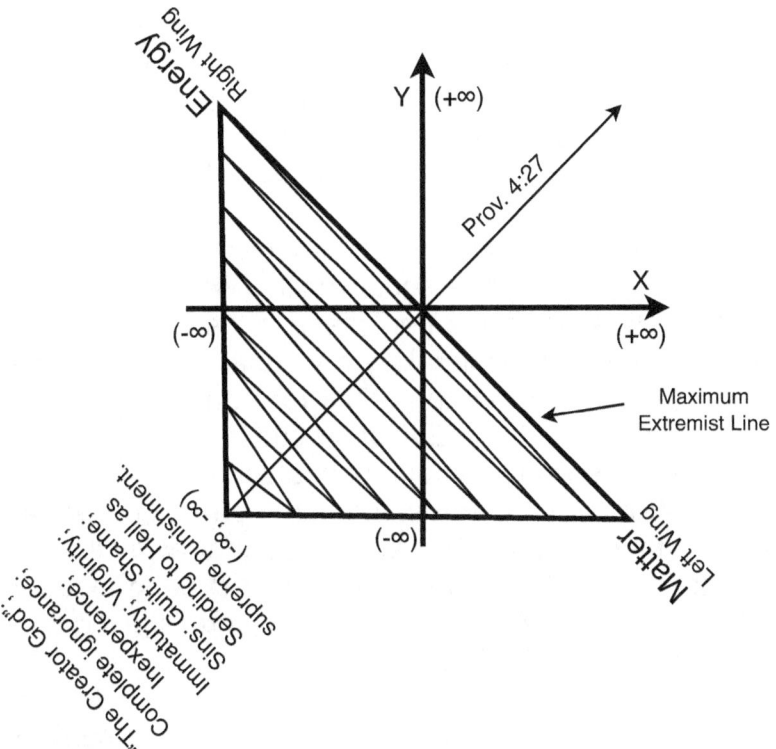

Inverted Holy Trinity Forces believe it was dark in the very beginning of it All, they believe the future will be white as it will be a better world. History of Mankind has shown us the Innovative Dark Forces were always the ones who brought innovation in mentality, in science, in education, even in those material goods for human daily consumption. History shows us how the "whites" has always been conservative, against evolution, having no arguments to beat the "Darks", they have always used force to pursue, condemn and kill, in the name of God, Allah, Odin, whatever. The first half of life, the underworld where souls pursue their basic needs, feel there are plenty of wrong doing in the world, but cannot really understand what it is. Things do not make any sense to the Believers, neither have to as they are the Inverted Holy Trinity Conservative White Forces.

Inverted Holy Trinity Forces believe it was dark in the very beginning of it All, believing the Future is white because of the light of knowledge and love it will bring. Their interpretation error stands, among other factors, in this: the Future will only become bright, and therefore white, when the light of our knowledge and love fall upon all over it, but our light of knowledge and our love, etc., will only start to fall upon the Future when the Future becomes Present, and will only completely fall upon it once it has became Past.

So, when does all the Graphic of Life becomes Past to one or a group of Souls? When these souls reach $(+\infty, +\infty)$! Meaning $(+\infty, +\infty)$ is the Heaven, is the only place in the Graphic of Life where there is total understanding and total love. For those whom stand in $(+\infty, +\infty)$, all the souls running the Graphic of Life stand behind them in the Evolution Latter, the Course Line of Life. For the souls who had not reached yet $(+\infty, +\infty)$, and are very far away from it, $(+\infty, +\infty)$ stands as the most distant and futuristic point there is in All the Universe, so far away they just can't tell what's going to happen then, it hasn't been made yet the light of knowledge for them on how the future is going to be, and this means... yes... the future is Dark!

This is the theory and the facts of Mankind History converge perfectly to it, as Mankind History has shown us that the innovative "Dark Forces" were always the ones that have brought innovation in mentality, in science, in education, in procedures, in products and services for All Mankind. History has showed us how the "White Forces" have been always conservative, against evolution, having no arguments to beat "Dark Forces", they have always used force to pursue, condemn and "kill" Dark Forces "in the name of God", Allah, Odin, whatever. The first half of life, the underworld where Souls pursue their basic needs, feel there are plenty of things "wrong", but cannot really understand what it is; they, who still live in the underworld U=]-∞, Maximum Extremist Line[, or U= {Inverted Holy Trinity}, are the ones who are going to "correct" by the use of force, all the (Un)Justice, all the "errors" the perfect God, Allah, Supreme Intelligence, the Great Architect of Universe,

22

(+∞, +∞), have allowed their sons to do. This is nothing but a genocidal mentality! Things don't have to make sense to them, as they are the Believers, the Inverted Holy Trinity Conservative White Forces.

Inverted Holy Trinity souls see the Universe upside down, as they truly believe in the "Creator God", who stands in the beginning of it All, meaning, in (-∞, -∞), in Complete Ignorance, in complete inexperience, in virginity, and stand for It. You can start by disagreeing this interpretation of the believer's God, Allah, Dja, Odin, the Great Architect of the Universe, whatever, by arguing an ignorant is completely unable to create anything, and a World, a Universe, a completely absurd idea. But is it truly so?

Can't a complete ignorant raise a family?

His kids will have no proper education, will have no comfortable place to live, will have no good looking, clean clothes nor clean teeth nr general health, his family will live in filth, in misery, living in sub-human conditions, like savages, not being able to understand anything around them, left all alone with their most basic needs of feeding, procreating and fighting for their power supremacy in a determined geographical area, but they will grow up that way, and the ones who might want a better life will have to work for it, an extremely difficult task to perform, since they will have to make a living from their own effort, and all alone, being expelled from their biological family's Common Sense whom believes that individual's difference is Evil and must be punished "Once and for All, to correct him".

Can't a complete ignorant create a business company?

Sure he can. The public office will ask him for the necessary documents, that he will get from his family, from some friend or some illegal move, not from his own skills. The public office will tell him where to sign and the ignorant becomes a business company manager.

If he will know how to run it properly, that's another issue, as the goal here is to prove any complete ignorant can become a manager, a "Creating God" in his community, occupying (-∞,-∞), the very beginning in the Graphic of its "corporation" lifetime.

So tell me...What did happen with all the living creatures, including Mankind, Homo Sapiens, in the very beginning of their lives? Did the Sons of God count with their father's help to get a good start? Or have the first living creatures grown up just like the sons of a complete ignorant? Didn't we All had to grow up in filth, in complete ignorance, didn't we All had to learn by ourselves, wasn't it because of that, that we all had, still have, and will have to, in the future, live and reincarnate several lifetimes? Weren't we all raised from the beginning as true sons of a complete ignorant? Still, that complete ignorant was the Creator One, It is the God, Allah, Dja, Odin, whatever, that in fact has created All in the beginning! So, it's in this God, Allah, Odin, whatever, Inverted Holy Trinity Souls believe in and stand for.

They stand for the solutions from the past, believing they can still be applied in the future with success. So they read every characteristic precisely opposite to Holy Trinity ones, meaning what H.T. reads at left wing, I.H.T. reads at right wing, and vice-versa, because they have an opposite point of view.

Headed to the future, to (+∞, +∞) and over its greater influence, Its goal is total knowledge and experience. The upper World, where there is no place for dogmatic beliefs but where understanding is a must, visionaries and pioneers. Here in this Upper World we don't make our stand with physical violence, but with knowledge, which is a spiritual characteristic. "The Kingdom of God", according to Jesus Christ's metaphors, meaning a place where we can understand it all, where we don't need to fear what we already know, because we have already learned how to face it successfully, and if it's not a threat, we can Love it All. For us, the beginning is already known, it's clear to us, so it's white. Also, "The Big Bang" was a huge explosion that spread light all over and it happened "in the very beginning" of what we know at the present time. What's going to happen in the future we don't know, it's not clear to us, so, the future is Dark. The Future will only become bright, and therefore white, when the light of our knowledge falls upon all over it, but our light of knowledge and our love, etc., will only start to fall upon the Future when the Future becomes Present, and will only completely fall upon it once it became Past.

So, as pioneers and innovators, we stand for the Dark, we are the Holy Trinity Innovative Dark Forces.

Holy Trinity Souls can understand The Graphic of Life, and so, they stand for the Ending God, Allah, Dja, Odin, the Supreme Intelligence, the Great Architect of the Universe, whatever, standing in $(+\infty, +\infty)$, right on the top of both the Evolution Axis and the Course Line of Life.

Every time we, Holy Trinity Forces say we stand for the Ending God, Inverted Holy Trinity Forces get in panic, shaking all up from their head to their feet, because for I.H.T. souls, standing for the Ending God is claiming for the definitive finishing of All the Life in the Universe, and since they can't even understand the Eternal Life and the Cycle of Life, for them, it's the End of the World, death, horror, Evil, we are standing for. So, they believe we are the "Dark Forces of Evil" and stand for killing us "Once and for All", like if that was possible, like if there weren't the Spiritual and Material Eternal Cycles of Life!!! Let's say for instance, we are on the degree 50%, matching the "Maximum Extremist Line". Some of us are ahead of it, others stand behind it, let's consider the difference is not that big, let's consider that Common Sense is now standing in U=]40%, 60%[. By the time the discussion around this book ends by being

finally accepted and approved, Human Society will enter another stage of evolution, witch we can define as U=]55%, 65%[, and then as U=]60%, 70%[and so on, until we can reach our goal of 100% possible Human evolution, understanding completely ourselves, the others, life and death, the Universe. Then, there will be nothing to fear, since $(+\infty, +\infty)$ is a mathematically open point, so Life does not end there. Achieving 100% of mankind evolution, and being $(+\infty, +\infty)$ a mathematically open point, means it will be the beginning of a new race, the next breed, the breed that will succeed Human kind, quite probably a new breed that will develop and adapt to an extraterrestrial ambiance. So we, the innovative Holy Trinity Forces, understand the Graphic of Life and we do stand for (+∞, +∞), the Ending God, Allah, Dja, Odin, the Supreme Intelligence, the Great Architect of the Universe, whatever, as our Dark King.

How can we objectively distinguish Holy Trinity souls from Inverted Holy Trinity ones? Before judging, pay attention to the fact that no one is 100% developed, basic, left or right wing, except (-∞, -∞), (+∞, +∞), extreme left wing where there is no human life but pure energy, and extreme right wing, where there is also no human life but pure matter. So, being a H.T. Soul, an I.H.T. Soul, a left wing soul or a right wing souls, can only mean that, that soul in particular stands in a point of view closer to (-∞, -∞), (+∞, +∞), extreme left wing or extreme right wing, in percentile terms. Once we have understood this, we can objectively identify Inverted Holy Trinity souls as the ones who, most of the times, stand for solutions applied successfully in the past, neglecting the entire atmosphere was different back then. Not that they are not conscious about it, it's just that they are believers, meaning their degree of evolution along the Course Line of Life does not allow them to understand, and Humanity has always been afraid of what it cannot understand, it's a natural defence mechanism of mankind, with the positive goal of saving unnecessary risks and death, as it is specified in the Cycle of Life. But like in many other issues, this mechanism is supposed to be used with moderation, because when we get a fundamentalist position over it, we are doing nothing more than closing (+∞, +∞)'s way to the future, to progress, both materially and spiritually speaking. They are still headed to (-∞, -∞), meaning to the past, they are conservative, what makes them Inverted Holy Trinity souls.

So, if Heaven, Hell and the Purgatory are stages, crossable by the Course Line of Life, it means Heaven, Hell and the

Purgatory, are degrees of evolution lived by the spirit as a Human Being as we know it, attached to its body/matter, as well as by the spirit in it's pure essence, dead from the human condition, although alive through its Cycle of Life (my book "The Eternal Life). So, this means the happiness and/or suffering we will live after death, is no different from the happiness and/or suffering we experience while alive in the human condition, because it is living in the human condition (in vegetable, animal and alien as well) that we have the free will to chose and the capacity to make things happen. That is why we need to reincarnate once and another time, because we do not have the chance to run all the Course Line of Life in one single lifetime, nor the capability to understand all the joy and the suffering we cause to others whom have crossed our lives, at any point, at any time. That's the way, the perfect God, Allah, Dja, Supreme Intelligence, Architect of the Universe, whatever, made each vegetable, each animal, the human being, the extraterrestrial life forms, all with its specific lifetime (expressed in years). Depending on the complexity of the degree of evolution to beat, each one needs its specific lifetime.

As Allan Kardec said in his "Book of the Spirits", "The Spirit sleeps on a rock, dreams on a vegetable, excites on the animal and wakes up in the Human Being".
Reading this, I ask a logical question: to do what, where and to be who?
Notice that no one wakes up just to be awakened! We all wake up every morning to do a specific job or to develop a specific study. So, I ask: "the spirit sleeps on a rock, dreams on a vegetable, excites on the animal and wakes up in the human being to be what? To do a specific job, witch is...?
"When" is a question that has to do with the degrees of evolution in the Evolution Ladder that I more specifically describe in this work as the Course Line of Life. Which will that step be? In which degree of evolution is Humanity standing now? Is that future degree of evolution we are talking about, the next one? Are we going to have some other in between?

So, one life the human being is a third quadrant soul (the true beginning), ignorant to produce material wealth or to understand its spiritual life;

The next Quadrant will be the Second, where he/she will feel tired of being ignorant about life, pursuing all the spiritual knowledge he can, living in the second quadrant. But he will be materially poor, will spread the good news about human spirituality but will have no power to feed all the hunger and to allow education to all mankind. These second quadrant souls cannot be happy here as well;

Next Quadrant will be the Fourth, where he will be able to create wealth to himself but will be still, very sarcastic, selfish, egocentric, as he'll be ignorant towards his spiritual essence, being materially rich but emotionally and spiritually unhappy.

Finally, it will ascend to First Quadrant, where it will be able to generate wealth and will have the spiritual knowledge and conscience to develop a philosophical, political, law, economic, social system, so that everyone can fly with its own wings, without having the ones with too much of an "A" and few of a "Z" persecuting to death all of those with too much of a "Z" and a few of an "A", and vice-versa. These souls will create wealth and lead all Humanity to do the same. Will be rationally altruist and will show by their own example to all mankind that, not only it is possible, but it is also the only way to go, where we can all be happy and share happiness with all Mankind.

Heaven, Hell and the Purgatory are made here, living in the vegetable, animal, human, extraterrestrial condition, are not places where we are going after our death. After death, we can only understand what went right and what went wrong, like it is described in the Spiritual Cycle, book "The Eternal Life". But we cannot do nor change anything there and that is why we must be reborn again and again, as many times as it takes. As the Course Line of Life comes from $(-\infty, -\infty)$ to $(+\infty, +\infty)$, we do have all the time in the world to make all the progress we need. As you can see, time is not a problem. Life forms will live as many lifetimes in each quadrant as they will need to, always regarding the Non-Retroactivity Universal Law. The Graphic of

Life allows you to use it forever by considering an Universe of the Present Time: U={Present Time}.

"Heaven", "Purgatory", and "Hell" are spiritual degrees of evolution. So, besides "left" and "right" wings, there's also "up" and "down". So how can we represent it mathematically?

By Inserting the Holy Trinity and the Inverted Holy Trinity in the Graphic of Life. You have now studied both the Holy Trinity and the Inverted Holy Trinity, so it's now time for me to explain the journey of our lifetime, as well as the journey of our eternal life.

You are now about to understand what is Heaven, Hell, the Purgatory, as well as another essential related things Humanity could not explain until now. You will now understand the sense of it, and realise that all is possible to understand, but when it is not, it is believed. Humanity has now reached a degree of evolution where it can change the "believing in what you cannot understand", by the "understanding of what you had once believed in". Free your mind from it all, concentrate, breath deeply, prepare yourself for the knowledge your mind is about to acquire and your spirit is about to remember. Once you have understood what I am about to explain, your vision and perception of the Humanity, the World and All the Universe, will never be the same. Beware, it is a point of no return.
Are you ready?
Please be honest with yourself... Are you really ready? When you are, please read my next book... "The Graphic of Life"!

Cheers brothers & sisters!